CSX Tunnel Fire
Baltimore, Maryland

Authored by: Hilary C. Styron

This is Report 140 of the Major Fires Investigation Project conducted by Varley-Campbell and Associates, Inc./TriData Corporation under contract EMW-97-CO-0506 to the U.S. Fire Administration, Federal Emergency Management Agency, and is available from the USFA Web page at http://www.usfa.dhs.gov

Homeland
Security

Department of Homeland Security
United States Fire Administration
National Fire Data Center

U.S. Fire Administration Fire Investigations Program

The United States Fire Administration develops reports on selected major fires throughout the country. The fires usually involve multiple deaths or a large loss of property. But the primary criterion for deciding to write a report is whether it will result in significant "lessons learned." In some cases these lessons bring to light new knowledge about fire--the effect of building construction or contents, human behavior in fire, etc. In other cases, the lessons are not new, but are serious enough to highlight once again because of another fire tragedy. In some cases, special reports are developed to discuss events, drills, or new technologies or tactics that are of interest to the fire service.

The reports are sent to fire magazines and are distributed at National and Regional fire meetings. The reports are available on request from USFA. Announcements of their availability are published widely in fire journals and newsletters.

This body of work provides detailed information on the nature of the fire problem for policymakers who must decide on allocations of resources between fire and other pressing problems, and within the fire service to improve codes and code enforcement, training, public fire education, building technology, and other related areas.

The Fire Administration, which has no regulatory authority, sends an experienced fire investigator into a community after a major incident only after having conferred with the local fire authorities to insure that USFA's assistance and presence would be supportive and would in no way interfere with any review of the incident they are themselves conducting. The intent is not to arrive during the event or even immediately after, but rather after the dust settles, so that a complete and objective review of all the important aspects of the incident can be made. Local authorities review USFA's report while it is in draft form. The USFA investigator or team is available to local authorities should they wish to request technical assistance for their own investigation.

For additional copies of this report write to the United States Fire Administration, 16825 South Seton Avenue, Emmitsburg, Maryland 21727 or via USFA WEB Page at http://www.usfa.dhs.gov

U.S. Fire Administration

Mission Statement

As an entity of the Department of Homeland Security, the mission of the USFA is to reduce life and economic losses due to fire and related emergencies, through leadership, advocacy, coordination, and support. We serve the Nation independently, in coordination with other Federal agencies, and in partnership with fire protection and emergency service communities. With a commitment to excellence, we provide public education, training, technology, and data initiatives.

ACKNOWLEDGMENTS

The United States Fire Administration greatly appreciates the cooperation received from the following people and organizations during the preparation of this report:

Donald Heinbuch	Operations Chief, Baltimore City Fire Department
George L. Winfield	Director, Public Works, City of Baltimore
The Honorable Martin O'Malley	Mayor, City of Baltimore
Jerry Young,	Public Works, City of Baltimore
John Nay	Chief Petty Officer, Field Response Operations Supervisor, United States Coast Guard
Scott Gorton	Safety Officer, CSX Transportation
Richard McCoy	Office of Disaster Control, City of Baltimore

TABLE OF CONTENTS

EXECUTIVE SUMMARY

At 3:07 p.m. on Wednesday, July 18, 2001, a CSX Transportation train derailed in the Howard Street Tunnel under the streets of Baltimore, Maryland. Complicating the scenario was the subsequent rupture in a 40-inch water main that ran directly above the tunnel. The flooding hampered extinguishing efforts, collapsed several city streets, knocked out electricity to about 1,200 Baltimore Gas and Electric customers, and flooded nearby buildings. The crash interrupted a major line associated with the Internet and an MCI WorldCom fiber optic telephone cable.

Throughout the incident, fire officials were plagued with three problems: fighting the fires in the tunnel; the presence of hazardous materials; and the weakening structural integrity of the tunnel and immediate surrounding areas.

Though the original cause of the fire is unknown, at the time of this report, fire officials believed that the derailment ruptured a tanker car carrying a flammable liquid chemical that fueled the fire. The fire response quickly jumped to five alarms within the first two hours of the incident. Thick black smoke emanated from both ends of the tunnel and seeped through manhole covers along Howard Street and other nearby streets.

Firefighters first attempted to fight the fire by entering the tunnel from either end, using vehicles with special rail wheels, but they were forced back by intense heat and a lack of visibility. They then initiated an alternative plan. Firefighters lowered large diameter hose from the street above into the tunnel where attack lines were set up for suppression operations. Firefighters were finally able to reach the burning cars at about 10:00 p.m.

The initial attack significantly lowered the temperature of the burning cars within a few hours.

At the height of the incident, 150 firefighters were on the scene working to extinguish the fire. For the first time since they were installed in 1952, civil defense sirens were activated at 5:45 p.m. to warn citizens of impending danger from the fire and hazardous materials. On the night of the derailment, city officials closed down entrances to the city from all major highways. Baseball games between the Baltimore Orioles and the Texas Rangers at nearby Camden Yards were postponed that night and the following night because of the hovering cloud of black smoke from the tunnel fire.

SUMMARY OF KEY ISSUES

Issue	Comments
Tunnel Access	Access to the 1.7 mile Howard Street Tunnel proved daunting to the emergency crews responding to the incident. Built in 1895, the tunnel was only accessible from either end and via a manhole located on Howard Street in the middle of downtown Baltimore.
Hazardous Materials	According to the waybills supplied by the CSX train crew to Incident Command, many of the freight cars in the 60-car train were carrying wood pulp and large rolls of paper combustibles. As reported, nine cars were carrying chemicals including five acid tank cars. Two tanks held fluorosilicic acid, two carried hydrochloric acid, and one held glacial acetic acid. Other materials on the manifest were ethyl hexyl phthalate (a plasticizer reducing embrittlement in plastics), tripropylene (a detergent raw material), and propylene glycol (an antifreeze ingredient).
Shipment Delays Along Route	The accident blocked CSX Transportation's only direct route between the industrial Northeast and the South, and delayed freight and passenger transportation. The Tropicana "juice train," which carries Florida orange juice to the New York-New Jersey market and is one of CSX's most time-sensitive shipments, was forced to detour through Harrisburg, PA, on a Norfolk Southern line.
Difficulties Encountered by Firefighters	Firefighters attempted to advance water cannons from each end of the tunnel in an effort to extinguish a fire they could not see. The heavy black smoke totally obscured vision inside the tunnel. Portable lighting was nearly useless. Firefighters attempting to enter the tunnel lost all vision within 300 feet of the entrance as the smoke deposited a black film on face pieces and goggles. The use of Self Contained Breathing Apparatus (SCBA) was essential. Gas masks and Air-Purifying Respirators (APRs) were useless.
Environmental Monitoring	Hazardous materials experts from the Maryland Department of the Environment tested the atmosphere repeatedly during the incident at both ends of the tunnel. Their reports did not include detection of acid content or other compounds of concern, but did include detection of significant wood-ash content in the heavy smoke from the burning combustibles freight. The United States Coast Guard also responded to the incident and provided air and water monitoring for hazardous materials. The majority of the air and water monitoring was conducted by a firm contracted by CSX. The Coast Guard conducted tag alongs with the company to verify the firm's air monitoring and water monitoring sampling results. The EPA conducted some air monitoring. Results for all air and water were unavailable at the time of this report, because of an ongoing investigation.
Heat Injuries	Two workers were treated for heat-related injuries from the fire that burned at almost 1,500 degrees Fahrenheit. They were released from the hospital after several hours.
Public Information and Media Accounts	Access to information was limited, because a public information officer (PIO) was not designated during the initial stages of the event, and also because CSX internal and external communications were problematic. Media accounts were sometimes vague and incorrect. This event demonstrated the importance of establishing a public information sector with a designated PIO as standard practice at all major events.
Delayed Notification	The time between the derailment at 3:07 p.m. and the time of the Baltimore Fire Department (BFD) notification allowed for the chemically heated fire to smolder, expand, and build up within the confined space of the tunnel. Delayed notification of the incident impacted the ability of the responders to gain access to the derailed cars and begin fire suppression activities and hazardous material assessment and containment. Delayed notification certainly impacted the financial cost of the incident to the City and CSX, as well as the number of personnel needed to respond to the incident. It is not known at this time whether delayed notification contributed to environmental impact, nor to what extent. Had there been immediate notification of the derailment, Fire Department personnel may have been able to contain the chemical spill and suppress the fire, thus reducing the arduous tasks and costs of a several-day response to the incident.

INTRODUCTION

A train enroute to Oak Island, New Jersey, from Hamlet, North Carolina, derailed in the Howard Street Tunnel under the streets of Baltimore, Maryland, July 18, 2001. The train was carrying a variety of freight and hazardous materials with three locomotives pulling 60 cars. Complicating the scenario was a subsequent rupture in a 40-inch water main that ran directly atop the tunnel. The flooding hampered extinguishing efforts, collapsed several city streets, knocked out electricity to about 1,200 Baltimore Gas and Electric customers, and flooded nearby buildings. The derailment also interrupted a major line associated with the Internet and an MCI fiber optic telephone cable.

The environmental impact from the five-day process of fighting fires and removing train cars is still being determined by investigators. The economic impact of the incident was felt by the City of Baltimore and downtown businesses. In October of 2001, CSX paid the City of Baltimore 1.3 million dollars to cover some of the costs of the derailment. The payment covered the cost of overtime for police, firefighters, and public works departments. It did not cover the cleanup of chemicals spilled, the investigation of the accident, replacement of the ruptured water main, nor road repairs associated with the derailment.

CSX officials accepted claims from 25 merchants on Howard Street, filed with the company's insurance adjuster, for the damage and lost business that resulted from the accident. In addition to the insurance claims, CSX paid $20,000 to a group representing businesses in the area that were closed following the derailment. In an effort to show good-faith and appreciation, CSX wrote three $5,000 checks to volunteer canteens that served meals to rescue crews responding to the train derailment.

THE HOWARD STREET TUNNEL

Background

Some of the earliest years of American railroading can be traced to the City of Baltimore with the founding of the Baltimore and Ohio (B&O) Railroad. Until 1884, the B&O railroad leased a railroad track through Baltimore to connect its eastern and western routes. In 1884, a competitor purchased the track, leaving the B&O with no way to get its trains through Baltimore. In an effort to salvage rail business through the city, the B&O worked with a contractor to construct a tunnel through the city from the Camden Station area north to the Mount Royal Station in the heart of Baltimore. The construction of the tunnel, formally called the Baltimore Belt Line, began in September in 1890 and was completed in May of 1895. Railroad historians believe the tunnel was the most expensive tunnel project of its time. The cost eventually drove the line into receivership in 1896.

A monument to railroad engineering of a hundred years ago, approximately thirty million bricks were required to build a rounded tunnel with a single track that rose at a steep grade—4.8 percent. The one-track line was originally 1.4 miles in length; an extension of three tenths of a mile was completed in the 1980s to accommodate parking for the major league baseball stadium and light rail construction. The depth of the tunnel varies from its shallowest at 3 feet to its deepest at 60 feet. The 4.8 percent grade accounts for a height difference of approximately 330 feet from the entrance to the exit at Mount Royal Station. The rail tunnel runs underground through such downtown landmarks as the Baltimore Arena, Maryland General Hospital, and antique shops along Howard Street. In 1958 the B&O railroad discontinued passenger service through the Howard Street Tunnel.

The tunnel remained little used for several decades. Today however, it is the longest active underground train route on the East Coast. According to a CSX Transportation representative, approximately 40 freight trains per day pass through the tunnel. Currently, the tunnel, rarely seen by residents, facilitates the passage of tons of freight, everything from orange juice to automobiles, fine goods, and coal. In September of 1987, the B&O railroad merged with and became CSX Transportation. In June of 1999 CSX transportation began operating a new rail network to include Conrail.

INCIDENT SCENARIO AND LOCATION

At the time the CSX train L41216 derailed in the Howard Street Tunnel, it was moving at 17 mph (eight miles under the posted speed limit of 25 mph.)[1] The train was halted by the emergency break, which was automatically applied. According to a representative of CSX Transportation, whenever a train's air brake system loses pressure, the brakes react automatically. The engineer cannot restart the train until the air sensor on the last car in the train acknowledges that it detects sufficient pressure.

The initial cause of the emergency break activation is still under investigation by the NTSB. However, it appears that the main air break hose, which runs the entire distance of the train, was either severed or disconnected. This can happen as the result of a malfunction, a decoupling, or a derailment. The train was stopped about a half mile from the north end of the tunnel.

CSX requires a two-person crew to operate a freight train. The engineer operates the controls of the diesel engine while the conductor watches for signals and is responsible for carrying the waybill-- an inventory of the cargo carried on board. When the train came to a stop, the engineer attempted to contact a CSX Transportation dispatcher by radio, but was unable to because the antenna would not transmit deep inside the tunnel. Unable to reach the CSX Transportation Dispatch Center at 3:15 p.m., the engineer used his cell phone to contact an official at the CSX Transportation dispatch center in Jacksonville, Florida. According to CSX Transportation, the protocol for response when the emergency brake activates is for the crew to dismount the locomotive and walk the length of the train in an attempt to locate the problem.

While in the tunnel, the crew did dismount the locomotives and began to walk, but they were met by heavy black smoke, no visibility, and conditions that made it difficult to breathe. The conductor and engineer were unable to find the derailed cars, and because of the untenable conditions, were forced back inside the train. The crew followed their Hazmat training and emergency protocols, shutting down the leading two locomotives, then uncoupling the third from the train cars so they could drive out of the tunnel and get away from the fumes and black smoke. According to the NTSB, sensors show the engines left the tunnel at 3:27 p.m.[2] Once outside the tunnel, the crew immediately radioed to their CSX employer, as required, and described the nature of the emergency as they understood it (heavy smoke and uncoupled train) and the steps they had taken.

Shortly after 4:00 p.m., the general public was beginning to notice a heavy black cloud pouring out of the tunnels and up through manholes. Calls started coming into the City of Baltimore Fire Department dispatch, reporting smoke. According to CSX Transportation officials, at 4:04 p.m., the

---

[1] Verified by the black box, according to the National Transportation Safety Board (NTSB).

[2] NTSB reports this time to be 3:25 p.m.

train crew contacted the CSX Transportation dispatcher in Jacksonville and Baltimore's 9-1-1 Center, reporting the emergency to each. While the public was notifying the Baltimore Fire Department (BFD), CSX Transportation dispatchers were still trying to get the exact location of the derailment pinpointed. A period of over one hour elapsed before CSX Transportation notified the BFD of the incident, thus confirming the call that was made earlier by the train crew.

The Baltimore City Fire Department responded to the Howard Street Tunnel and upon arrival at 4:18 p.m., the Freight Conductor supplied the "Rail Waybill," including the "Hazmat Chart," to the on-scene Incident Commander. The waybill identified all the cars numbered in order from car one (immediately behind the last locomotive) to the last car in the train, allowing emergency responders to know the exact location of each car and what it was carrying. Between the rail waybills and the Hazmat Charts, the Incident Commander had a computer printout providing full details of all the cars and identifying those that contained hazardous materials.

According to the Waybill, 31 cars were loaded and 29 cars were empty. A line-order and location of cars and their contents are as follows: three locomotives, four cars of pulp board, 17 empty cars, one car of pulp board, one empty car, one tank car of propylene glycol, one tank car of glacial acetic acid, one car of flat steel, two empty cars, two tank cars of fluorosilicic acid, one empty car, one car of bricks, four tank cars of soy oil, four empty cars, eleven cars loaded with paper, one tank car of tripropylene, two tank cars of hydrochloric acid, one tank car of ethyl hexyl phthalate, one car of pulp board, and four empty cars. Hazardous materials present in the cars are described below in Table 1 as reported by CSX Transportation officials.

Table 1. Hazardous Materials Present In The Cars[3]

Hydrochloric acid	A metal cleaner. Not combustible, but highly corrosive. If inhaled, can cause a burning sensation, cough, labored breathing, shortness of breath and sore throat. On contact with skin or eyes, can cause severe burns.
Glacial acetic acid	A glass solvent. Flammable. If inhaled, can cause sore throat, cough, burning sensation, headache, dizziness, shortness of breath, or labored breathing. On contact with skin or eyes, can cause pain, redness and severe burns.
Fluorosilicic or hydrofluoric acid	Used to fluoridate water. Not combustible, but corrosive. If inhaled, can cause a burning sensation, cough, and shortness of breath. On contact with skin or eyes, can cause pain, redness, blisters and severe burns.
Propylene glycol	A de-icing fluid. Combustible. Can cause redness and pain in eyes.
Tripropylene	A lubricant similar to paint thinner.
Ethyl hexyl phthalate	Used to make flexible products like PVC piping. Combustible. If inhaled, can cause cough or sore throat.

[3] Figure 1: *Baltimore Sun Newspaper.* July 19. 2001

FIREGROUND OPERATIONS

Alarm Sequence

Table 2 below shows a breakdown of the alarm sequence as received by the Baltimore Fire Department and their dispatch. The documentation of times and the subsequent striking of additional alarms are according to the official BFD dispatch record of this incident and were provided by Donald Heinbuch of the BFD. It should be noted that the first response was a tactical box alarm with Battalion Chief (less than a full box assignment), which would consist of four engines, two trucks, and a Battalion Chief. The tactical box is a standard response to a "report of smoke," the type of call as reported.

Table 2. Howard Tunnel Incident Alarm Sequence

Time	Description
4:15 p.m.	Tactical Box with Chief E13, E6, T16, BC2
4.18 p.m.	E13 takes command
4:23 p.m.	When E13 found out that Hazmat was involved, he called for the remainder of the Box which was E102, E8, TI
4:23 p.m.	BC 6 (Hazmat CHIEF), hearing the EI 3 report Hazmat of a potential condition, ordered communications to strike a full Hazmat BOX, which was added to E13 request for the remainder of the Box Assignment. E58, E57, E35, T21, Hazmat 1, Medic 9, Rescue 1, BC6 added to first alarm
4: 25 p.m.	BC2 establishes (assumed) command
4:31 p.m.	BC2 asks for second alarm - E31, 33, 10, BC3
4:43 p.m.	Shift Commander assumed command
4:48 p.m.	Acting Assistant Chief of Planning establishes and takes charge of Camden sector (south)
5:07 p.m.	Acting Assistant Chief of Operations command orders 3rd & 4th alarm to Mt. Royal sector (north)
5:08 p.m.	Third alarm E52, 36, 14, T26, E124 (TOWER)
5:11 p.m.	Fourth alarm E2, 29, TS, 15
5:13 p.m.	Fifth alarm ordered to Camden sector
5:13 p.m.	Fifth alarm E26, 21 T23
6:25 p.m.	Water main break reported / smoke change
After 6:30 p.m.	Several attempts to enter from north end
After 6:30 p.m.	Smoke tested by MDE/Hazmat 95% steam 5% ordinary combustibles
9:00 p.m.	Plan to enter tunnel from south end
9:50 p.m.	Entry into south end
10:12 p.m.	Reentry to collect sample of runoff 7/19
12:30 a.m.	Planning to relocate command to south end (Camden sector)

At 4:15 p.m., an alarm was sounded for smoke coming from the train tunnel at the Mount Royal Station. Engine Company 13, seven blocks away, had responded to the Mount Royal Station on many previous occasions for reports of smoke in the vicinity. Usually someone mistook the diesel engine smoke exiting the tunnel for a fire. However, the Captain of Engine 13 knew from the volume of black smoke emanating from the tunnel that this time the situation was different. At 4:23 p.m., after conferring with the CSX engineer and conductor over the waybill, the first due engine company officer requested that the Hazmat team respond. Hearing El 3 reporting a potential hazardous materials incident, BC6 ordered dispatch to deploy a full Hazmat Box. With BC2 in command, additional alarms were requested to augment specialized resources, such as The Maryland Department of the Environment and the U.S. Coast Guard, filling out the resources at either end of the tunnel.

Tactical Response

The first arriving companies and command staff instituted the incident command system as prescribed by the Fire Department's standard operating procedures. Based on the initial size-up and information regarding the train and its hazardous cargo, a plan of action was adopted. To ascertain the condition, extent, and progress of the fire, the Engine 13 Captain and approximately a half-dozen firefighters wearing standard turnout gear went to the southern end of the tunnel through smoke and intense heat with hoses, trying to reach the train that was sitting about three-quarters of a mile to the north. The team got within about 300 yards of the cars, but had to retreat because of intense heat and heavy smoke.

The initial attack to control the burning cars in the tunnel was futile. Approximately three hours after the derailment, at 6:25 p.m., a report of a water main break on Howard Street above the tunnel was received. A cooperative decision between Incident Command (IC) and the Baltimore Public Works Department (DPW) was made to allow the water to flow from the forty-inch ruptured main into the tunnel for approximately two hours. With no firefighter access to the derailed cars, the potential for a Boiling Liquid Expanding Vapor Explosion (BLEVE) was reduced by allowing the water to flow. A Public Work's official explained that approximately 60 million gallons of water flowed from the City's water supply during the two-hour time period. The smoke dramatically changed in color— from dense black to light gray then to white—indicating that the flowing water helped extinguish the fire.

The Maryland Department of the Environment (MDE) and Hazmat teams tested the smoke exiting the tunnel, and indications were that the smoke consisted of 95 percent steam and 5 percent ordinary products of combustion, indicating that the broken water main was indeed having a positive effect. The MDE determined that air tested within the affected outside areas near the tunnel portals was not toxic. Testing did reveal the presence of wood ash within the smoke, possibly caused by burning crossties as a byproduct of the fire.

Temperatures in the tunnel reportedly reached approximately 1,500 degrees Fahrenheit causing the freight cars to glow like steel in a blast furnace. Because of the conditions, firefighters were unable to predict when they would control and extinguish the fire, let alone remove the cars and the hazardous cargo from the tunnel. Unable to reach or extinguish the fire from either end of the tunnel, an alternative firefighting plan was initiated by noon of the second day. The decision was made to enter the only other known portal, a manhole entrance at Howard and Lombard Streets. At this point a five-inch diameter hose was lowered into the manhole and firefighters simultaneously entered the southern entrance of the tunnel to find and connect their hose lines to the five-inch water supply.

Over the next several hours, firefighters were able to significantly lower the temperature in the tunnel and help bring the fire under control.

With this tactic in place, firefighters could approach the still-burning cars and investigators could begin their investigation at close range. By 0900 hours Thursday, five cars from the end of the train were removed after many hours of struggle. The fifth car of this group contained a load of pulp board, which required several hours to extinguish with the help of a backhoe.

Hazardous Conditions Restricting Firefighter Intervention

Any incident involving the potential for vaporized hazardous materials, including the possibility of acid vapor splashing[4], requires full body protection in the form of Level A chemical protective clothing. That is quite different from standard firefighter 'turnout gear'. With the combined requirements for flame protection, supplied air respiration, and protection from chemical contact, and facing the hazards of confinement in an enclosed space with the potential for a BLEVE imminent, and vision totally obscured--firefighters were presented with a situation too dangerous to approach. Some of the hazard factors had to be reduced over a period of many hours before firefighters could enter the tunnel safely.

Chemical hazards were present for at least three days. Car number 53, containing 20,000 gallons of hydrochloric acid, lost nearly one quarter of its load. The spill and seepage crept into the storm sewer and was carried to the Inner Harbor. A member of the Maryland Department of the Environment hazardous materials team said the car was leaking at the seams. Fixing the leak of tank car 53 was complicated because the adjacent tank car (number 52), containing tripropylene, was completely burnt. Investigators believe that the tripropylene tank fire fueled the tunnel fire and made heat so intense that it was almost impossible to remove the hazardous chemicals.

Ironically, reports from the *Baltimore Sun* state that the Maryland Department of the Environment sampling found no hazardous compounds in the smoke coming from the tunnel; however, tripropylene produces irritating, but not highly toxic fumes when it burns. Because of the blaze, firefighters could not neutralize the acid leak by spreading an alkaline chemical to neutralize it. In the presence of such intense heat, the combination would have been explosive. Traditional plastic pumps would also not have survived the heat. To cool the acid, firefighters opened the manhole cover and sprayed the leaking car with water for over two hours.

On the third day of the incident, CSX contractors began pumping hydrochloric acid from car 53 and the adjacent car 54, which was not leaking. A vacuum hose was lowered through the same manhole that was being used to allow access into the tunnel for firefighting purposes. Additional contractors hired by CSX began replacing the 800 feet of damaged track at the south end of the tunnel to expedite removal of the remaining cars.

---

[4]Acid vapor splashing may occur as a result of vapors escaping from their containment vessel. Vapors rising to the ceiling of the tunnel and having no route to escape would then cool in the atmosphere, causing condensation that may result in splashing, dripping, and pooling.

FIRE COMMUNICATIONS

Fire Communications Bureau

The City of Baltimore has a consolidated Communications Center with an 800 MHz Fire, Police, and Public Services frequency system. In 1995, the City of Baltimore awarded a contract to design and build a state-of-the-art radio and Computer Aided Dispatch (CAD) system. Phase I involved the construction of nine antenna sites connected by a fiber optic SONET ring and the installation of the Fire Department radios and CAD. Completed in July of 1998, the system has given outstanding performance to the Fire Service during major emergencies, including the Howard Street Tunnel.

The Incident Commander believed that communications and radio capability of the BFD were the most important features of the incident's success. The second success factor was interagency cooperation. The BFD radio systems includes

- Mapping display of emergency locations

- System redundancy to ensure maximum reliability

- Single call for service to generate multi-agency response

- Inter-agency communications at major emergencies

- Automated service calls for better tracking and follow-up

- Shared resources, information, facilities, equipment, and costs

- Enhanced coverage in buildings, tunnels, and all areas of the city

- Emergency button for life threatening situations

- Automatic Vehicle Locators (AVL) on Medic Units

Resource Response

The Baltimore City Fire Department carried out a noteworthy response in its tactics to contain, manage, and resolve the tunnel fire incident. As the events of the afternoon of July 18, 2001, unfolded and time elapsed, it was evident that the incident would require an extended effort. At the beginning of the incident, the amount of time and resources were unknown. Eventually, more than 100 hours spanning four days would be spent bringing this incident under control. At the height of the incident, over 150 firefighters from both the City of Baltimore and Baltimore County were on the scene.

City, state, and federal government agencies deployed resources to the command post as follows:

- Baltimore City Fire Department

- Baltimore City Police Department

- Baltimore City Emergency Management

- City of Baltimore Department of Public Works

- CSX Transportation

- Baltimore County Fire Department

- Representatives from the South Baltimore Industrial Mutual Aid Plan were at the scene with personnel that include chemists and other hazardous materials specialists

- Maryland Department of the Environment

- National Transportation Safety Board

- United States Coast Guard

The success of this incident is directly related to the interagency cooperation and coordination of agencies and resources. The responding agencies played a part in providing the logistics, equipment, supplies, and expertise that controlled and limited the damage to the property, environment, health, and welfare of the citizens of Baltimore.

The United States Coast Guard deployed a series of floating booms to protect the Inner Harbor against contamination and potential hazardous runoff from the derailment site. It was feared that some of the drainage from the tunnel would find its way into the Inner Harbor.

EMERGENCY PREPAREDNESS

Baltimore fire officials recently had conducted a drill in one of the city's Amtrak tunnels using a MARC train; they also ran drills in a Metro tunnel. The drills were intended as training exercises in the event of a passenger train accident--not one involving a freight train--but they did acquaint the personnel with the environment of a railroad tunnel.

According to the City of Baltimore Emergency Management Plan, a public information announcement is required to be given during a Level II emergency.[5] In the early stages of the incident, the incident was determined to be a Level III emergency and the Emergency Management Director urged that a public announcement be made over radio and television to alert citizens and to initiate a shelter-in-place advisory that citizens stay inside their homes until the chemical hazards could be assessed. In the general area surrounding Mount Royal Station, citizens were offered the choice to leave or shelter-in-place.

For the first time since installing them in 1952, the city activated its civil defense sirens at 1745 hours to warn citizens of the impending danger from the derailment and fires. Since there was some concern over the residual effects of smoke to persons around the tunnel portals, an evacuation order to pedestrians was broadcast. Persons living near the affected areas were told to stay indoors, seal up windows, and turn off air conditioners.

POLITICAL CONSIDERATIONS AND INVOLVEMENT

The train derailment in Baltimore focused attention once again on the issue of transporting hazardous material, including radioactive and nuclear waste, through densely populated areas. The cargo of the CSX L41216 included hydrochloric acid, 5,000 gallons of which spilled before workers began pumping it out. The train also carried tripropylene, a combustible lubricant similar to paint thinner,

---

[5] Public action is not necessary for a Level I emergency. Level II emergency states that planning for public action should be considered and a public information announcement should be made. Level III Emergency states that public action is necessary and that the public should be made aware of the situation and ordered to evacuate or shelter-in-place.

and hydrofluoric acid, a corrosive used in making gasoline. The derailment was unusual; the shipment was not. CSX reports that 40 freight trains run through Baltimore on an average day. Some days, all of them carry hazardous materials. Trainloads of hazardous materials cut through metropolitan areas around the country on a daily basis. With the transport of hazardous materials, it is crucial to have the involvement of a Local Emergency Planning Committee (LEPC) and the Fire Department to provide community awareness and interagency cooperation.

The U.S. Nuclear Regulatory Commission design criteria for high-level nuclear waste containers calls for casks to be able to withstand a 1,475 degree fire for 30 minutes. The Baltimore tunnel fire temperatures, estimated to be greater than 1,500 degrees, surpassed the NRC's design criteria for containers that would hold atomic waste. One Senator has complained that the NRC criteria go back to 1947 without any updates, despite combustibles on the roads and rails today that burn at much higher temperatures.

On October 5, 2001, at a United States House of Representatives Subcommittee on Government Efficiency, Financial Management, and Intergovernmental Relations hearing, Baltimore's Mayor testified on the readiness, preparedness, and vulnerability of the City of Baltimore. The full testimony may be viewed in Appendix A.

> "...Baltimore had a chance to test our readiness in a chemical incident this past July, when a CSX train loaded with toxic chemicals derailed and burst into flame—burning in a tunnel beneath our city for five days....
>
> During the train fire, as is the case in virtually any crisis, local government was the first on the scene. Baltimore's Fire Department arrived within minutes after being contacted by CSX The Police Department, coordinating with the Fire Department and our Transportation Office, began to reroute traffic and secure areas that presented potential dangers—including Camden Yards, only a few hundred feet away from the tunnel, where the second game of an Orioles doubleheader was scheduled to begin. And the Health Department was monitoring air quality....
>
> One thing that was immediately apparent was the need for an effective incident command structure. [emphasis added] Right away, we knew the accident was serious, but not how serious. Within hours, we had the convergence of every level of government: Baltimore's Fire, Police and Health Departments; the State Departments of Transportation, and the Environment; and the National Transportation Safety Board....
>
> Given the rapidly changing state of information—which I believe is the case during any emergency situation—we needed to integrate all of these government agencies into a command structure capable of quickly receiving, evaluating, acting upon and disseminating information.
>
> The fire was the most immediate problem that needed to be addressed, so our Fire Department assumed control of the accident scene until it was extinguished several days later. The Governor's order that the State agencies to defer to local decision makers was critical in making this operation run smoothly.
>
> The CSX fire had one more complicating factor, which was probably not atypical of a potential terrorist attack. We were forced to work in multiple locations: Camden Yards at the south end of the tunnel, Mt. Royal at the north end, and a manhole in the middle of downtown that was directly above the burning train.
>
> One of the first things we realized—based on our experience in the CSX tunnel fire—was that most rail yards and tracks, filled with chemical tanker and munitions cars, represent one of our most vulnerable targets....
>
> Baltimore is like every other city on the East Coast in this vulnerability. Hazardous materials are shipped by rail within yards of residential neighborhoods. There were large apartment buildings right next to the tunnel where our train fire burned Commerce cannot stop, but this is one area where the federal government can make a significant impact."

ECONOMIC IMPACT

CSX Transportation

The cost of the train derailment and fire remains undetermined. CSX Transportation faces millions of dollars of clean-up costs and reimbursement payments to the State of Maryland and the City of Baltimore. On July 25, 2001, during a meeting at Baltimore's City Hall, CSX agreed to pay the overtime costs for the fire, police and public works forces related to the fire. CSX indicated, however, that the payment should not be construed as an acknowledgement of blame or fault. The cause of the derailment is still under investigation. The city initially estimated the clean-up to cost about $1.3 million.

Meanwhile, CSX canvassed the business establishments within the affected area to solicit claims for business losses resulting from the emergency. With the cause of the derailment still under investigation, CSX ran a full-page advertisement in the Baltimore Sun, entitled "Thanks, Baltimore!" The advertisement, addressed to the citizens of Baltimore, thanked the mayor, fire chief, "the courageous professionals of the Baltimore City Fire Department," and the emergency response personnel for their "tireless efforts, leadership and professionalism" following the derailment, it also thanked the community for its patience and support.

Major League Baseball Impacted

The Baltimore Orioles baseball team also was affected by the train derailment and fire. The rising black smoke coming from the tunnel traveled toward nearby Camden Yards Stadium—home of the Baltimore Orioles—where a game was in process. Public safety officials were forced to evacuate the stadium where the Baltimore Orioles were hosting a double-header day game. Ultimately, the Orioles postponed four games in three days because of public safety concerns, estimating the loss at $4.5 million. By Saturday, July 21, the baseball game between the Anaheim Angels and the Baltimore Orioles was played in spite of the fact that Howard Street was still closed between Mount Royal Avenue and Pratt Street. Interestingly, that game was "Fire Fighters' Appreciation Night," a commemoration that had been planned long before the derailment occurred.

General Business

The City of Baltimore issued a liberal leave policy for employees, as did the State of Maryland for its employees at State Center. As utility workers, police, and firefighters labored on Howard Street, two large office buildings and several other businesses along the road were shut down. The downtown business district and Howard Street especially, took a huge economic hit from the train fire. The director of economics for the Maryland Business Research Partnership stated, "You have to worry about Howard Street. If the damage is big enough, businesses could be closed for months. They suffered through light rail being built and hard economic times already."

Fiber Optic Networks Affected

When the CSX Transportation freight train derailed and caught fire, fiber optic cables serving several large carriers were damaged. The Director of Operations for Metromedia Fiber Network (MFN), located in Baltimore, noted that it was immediately apparent that they would not be able to access the cables in the tunnel anytime soon, given the situation with the fire and hazardous materials.

Although backup systems kicked in immediately, MFN was determined to restore full redundancy to its network as quickly as possible.

The fiber optic network company, city officials, and a major telecommunications company worked together to manage the process of rerouting cable through a maze of conduits accessible by manholes beneath the center of the city. Although a large portion of the recovery route already had been set up for possible use, workers had to clear blockages in at least four locations—a task that included digging up the street and rerouting the line. To restore full redundancy to its network, the fiber optic network company deployed emergency crews to re-route 24,000 feet of fiber through a maze of city conduit in 36 hours.

A construction services company completed a new fiber optic loop around the downtown tunnel where the derailment had damaged cables that transport Internet data, e-mail, and phone calls. Crews had nearly completed installing 30,000 feet of fiber when they were stymied by thick mud that was discovered in underground ducts along Light Street.

Media reports stated that a Silicon Valley company tracking Internet traffic said the train accident caused the worst congestion in cyberspace in the three years that it has monitored such data. The link through Baltimore "is basically the 1-95 of Internet traffic into and out of Washington," said the Director of Public Services for a company that monitors Internet flow by the hour on its Web site. The accident had almost no impact in some areas, including parts of Baltimore, while certain connections were 10 times slower than normal, such as the ones between Washington, D.C., and San Diego.

TRANSPORTATION AND PUBLIC WORKS LOGISTICS

The Department of Public Works (DPW) sent its hazardous materials specialist to the Command Center to coordinate with Incident Command. The DPW was responsible for controlling traffic in the downtown area and coordinating with the Maryland Transportation Authority and the State's Department of Transportation. All traffic coming from the downtown area was re-routed away from potentially dangerous areas, and traffic on highways inbound to the city was stopped and rerouted. The changes in traffic routes caused rush hour gridlocks and affected even the light rail transportation in the city.

While rush hour traffic was disrupted because of the fire, the Camden Line service was curtailed where bus service was substituted. Baltimore's light-rail line, which runs adjacent to the CSX line at both ends of the affected tunnel and operates on Howard Street directly above the tunnel, had to be suspended. Amtrak Northeast Corridor and MARC Penn Line services were not affected, although Baltimore's Penn Station is only about three blocks from the east portal of the CSX tunnel.

Two days into the incident, a domestic and international mailing company reported on the shipping delays caused by the CSX train derailment, indicating that dozens of trains were rerouted and some remained idle while the fire continued to burn inside the tunnel. Rail traffic headed south from the Port of Baltimore was detoured for hundreds of miles, and officials announced that rail shipments might be delayed as long as two weeks.

The DPW estimated that approximately 60 million gallons of water from the City's water supply were used to assist fire suppression activities. The water supply to the Shock Trauma Hospital had to be maintained and the Water Department had to re-route several pipe channels to ensure water service to the hospital and other businesses.

By 6:00 p.m. on Sunday, July 22, the remaining train cars were being removed; initial repairs to the broken water main and to a collapsed storm drain had begun. To access the affected water main and storm drain, a portion of the light rail line's track had to be cut. The necessary repair work caused further delays in the resumption of light rail service along Howard Street above the tunnel.

By Monday, July 23, the final car had been removed from the tunnel. Shortly thereafter, work began to assess the structural integrity of the tunnel and to continue repairing both the broken water main and the collapsed storm drain. Traffic remained grid-locked, however, because of the inability to cross Howard Street along the mile-long north-south corridor.

NTSB INVESTIGATION

The National Transportation Safety Board (NTSB) estimates that it could require more than a year to conclude the investigation and to report its findings regarding the cause and circumstances of this accident. Meanwhile, NTSB is working with the city and the railroad in an attempt to pinpoint the cause. One of the questions that is being addressed pertains to the circumstances surrounding the water main break at Howard and Lombard Streets three hours into the incident.

The NTSB is studying records of water flow and pressure supplied by the city in an effort to determine exactly when the affected system failed. The NTSB cautions that this is just one theory among others being considered, and that it may be several months or longer before there is a conclusion. NTSB took samples of the rail and water main for metallurgical analysis as part of its ongoing investigation over the cause of the accident. They are also reviewing the city's Water Department maintenance records for the 40-inch main that cracked.

Questions have also arisen over the amount of time between the accident and notification to authorities at the onset of the fire. The accident is said to have occurred at about 1507 hours, but fire officials say they did not learn about the fire in the tunnel until about an hour afterward. The delay in notification certainly impacted the severity of the incident, which grew for over one hour in the confined space of the tunnel.

INCIDENT INJURIES

Two workers on the tunnel incident were hospitalized after trying to inspect the derailed train. The workers were treated for heat-related injuries from the fire that burned at about 1,500 degrees Fahrenheit. They were released from the hospital after several hours. Four emergency workers, two of whom were from CSX, were rescued by an engine company from the eastern end of the tunnel when at least one of the workers complained that his supply of oxygen was running out.

Baltimore Division General Manager of CSX Transportation reported that this was the worst train incident he had ever experienced, describing it as "just short" of a worst-case scenario. He added that the "buddy system" was in place for workers within the tunnel, usually working in groups of seven or more.

LESSONS LEARNED

Issue	Comments
Media Relations	Designate a Public Information Officer for the duration of the incident. This incident occurred in the afternoon, and by the 5:00 p.m. newscasts, a PIO was still not available. According to Incident Command, inaccurate reports were made regarding the seriousness of the situation.
Maintain Control of "helpers"	Incident Command must maintain control of mutual aid and volunteer resources responding to incident. A few times during the initial incident coordination, volunteer response and representatives from City services left the IC staging area and had to be located for certain tasks to be accomplished.
Maintain Control of Workers	Accountability for personal safety and safe operations is a must during incident response and duration. Incident Command was able to maintain control and accountability of firefighters responding to the scene. As a result, there were no injuries to personnel (other than heat-related) and manpower resources were properly maintained to meet the needs of the incident.
Tactical Response and SOP's	City management should review and update strategy and SOP tactics for response to tunnel incidents. IC recommends that other cities with a railroad industry collaborate to be successful and comprehensive.
Training and Drills	Expand use of drills and training along with the relationships with the partners involved in response. The value of training and drills cannot be expressed enough. The "what ifs" should include the worst-case scenarios and most complicated response needs. The BFD believed they would remain prepared to respond to incidents of this type and scale with continued training and drills for the entire infrastructure.
Partnership	Develop strong working partnerships with agencies and businesses responsible for responding to incidents. The BFD acknowledges that the incident response was successful because they had established working relationships with other City resources and businesses and industries in their area.
Incident Management System	Expanding on the partnership lesson, this incident demonstrates the importance of training and pre-planning. By instituting pre-defined plans for response that were developed with flexibility to expand during an incident, the partners were able to reach consensus on appropriate response tactics and ultimately expand the operations into a Unified Command Structure.
Emergency Management Plan	The BFD noted that the City's emergency management plan needed to be reviewed and revised to incorporate technologies, safety, health and environmental impact, and other issues. Major incidents in any city present an opportunity to test under actual circumstances what works well, and what needs improvement insofar as emergency management plans are concerned.
No Serious Injuries	Adherence to Incident Command decisions and cooperation, contributed greatly to a low number of injuries—only two, which were heat-related.

CONCLUSION

The investigation as to the cause of the CSX train derailment is still pending at the time of this report. One conclusion that can be drawn, however, is that the response, fire suppression, and containment activities were successful in part because of previous disaster response planning and field exercises. Mutual aid agreements and a designated command structure provided the framework for coordination and cooperation among city and state officials. Practicing response exercises involving unified command and multi-agency response paid off in measurable ways. The CSX train derailment put the City of Baltimore's Emergency Response Plan to the test, and overall, officials were pleased with the response. While the lessons learned from this incident provided the City with critical information on where the Plan could be improved, officials saw first-hand the value of their on going efforts in emergency response planning. At a time when cities across the country are considering the potential impact of a terrorist incident, Baltimore's successful example of comprehensive pre-disaster planning points to why such activity is so essential, particularly when it is structured around the incident command system. The threat of incidents involving weapons of mass impact—chemical, radiological, biological, nuclear, and high-yield explosives—places higher demands on government to develop and test strategies and tactics for commanding multiple agencies during complex incidents over extended periods of time. Baltimore has taken major steps toward accomplishing this goal.

APPENDICES

APPENDIX A

MARTIN O'MALLEY
Mayor
250 City Hall
Baltimore, Maryland 21202

October 5, 2001

Testimony of Baltimore Mayor Martin O'Malley

SUBCOMMITTEE ON GOVERNMENT EFFICIENCY, FINANCIAL MANAGEMENT

& Intergovernmental Relations

Committee On Government Reform

Mr. Chairman. Members of the Subcommittee. Thank you for the opportunity to join you today. In acquainting myself with the Members of the Committee, I discovered that many of you bring unique insights to our nation's preparedness efforts.

In this time, we all have our crosses to bear — some heavier than others. But they will be lighter if we bear them together. I am eager to hear your insights into how we can better prepare for what we now know is a very real threat.

In many ways, Baltimore is a typical large city in terms of what we must do to protect our citizens from potential terrorist attacks. We have a few high-profile targets. We are not the largest city, but we're not the smallest city. And we take these issues far more seriously than we did before the tragedy of September 11th.

However, in some ways, Baltimore was in a unique position to come up to speed quickly on this issue. And we are sharing our experience with the US Conference of Mayors. On Tuesday, we will hold our second teleconference, sponsored by the Conference of Mayors and our President Marc Morial of New Orleans. The first dealt with biological weapons, and the second will deal with chemical weapons.

Some of you may know, we were selected as a lead city in the Chemical Warfare Improved Response Program, due to our proximity both to Washington, DC and the US Army Soldier and Biological Chemical Command (SBCCOM) in Aberdeen, Maryland. Baltimore also is home to the Center for Civilian Biodefense Studies at Johns Hopkins University — the only institution of its kind in the country.

We are fortunate to have a Police Commissioner, Edward Norris — from whom you will hear shortly — who, as a Deputy Police Commissioner, was involved in New York City's civil preparedness efforts in the wake of the first World Trade Center Bombing. Commissioner Norris has a nationwide network of law enforcement experts on which we have been able to draw.

Finally, Baltimore had a chance to test our readiness in a chemical incident this past July, when a CSX train loaded with toxic chemicals derailed and burst into flames — burning in a tunnel beneath our city for five days. The train fire is where I'll start.

During the train fire, as is the case in virtually any crisis, local government was the first on the scene. Baltimore's Fire Department arrived within minutes after being contacted by CSX. The Police Department, coordinating with the Fire Department and our Transportation Office, began to reroute traffic and secure areas that presented potential dangers — including Camden Yards, only a few hundred feet from the tunnel, where the second game of an Orioles doubleheader was scheduled to begin. And the Health Department began monitoring air quality.

One thing that was immediately apparent was the need for an effective incident command structure. Right away, we knew the accident was serious, but not how serious. Within hours, we had the convergence of every level of government: Baltimore's Fire, Police and Health Departments; the State Departments of Transportation, and the Environment; and the National Transportation Safety Board.

Given the rapidly changing state of information — which I believe is the case during any emergency situation — we needed to integrate all of these government agencies into a command structure capable of quickly receiving, evaluating, acting upon and disseminating information.

The fire was the most immediate problem that needed to be addressed, so our Fire Department assumed control of the accident scene until it was extinguished several days later. The Governor's order that State agencies to defer to local decision makers was critical in making this operation run smoothly.

Crisis breeds confusion, when what you need, right away, is clarity. The only way to quickly achieve clarity is to prepare — making as many decisions as possible in advance. In the event of an emergency, each level of government — and each agency — must prepare, coordinate, respond and adjust. They should know:

- Who are your critical personnel?

- Where should you assemble, with your peers from other agencies, to establish an initial command center?

- How can you remain in contact through an effective, redundant communications infrastructure — including cell phones, pagers, two-way radios and Blackberrys, in addition to landlines? You never know when your primary system might go down.

- Where will you meet to brief the press, providing instructions to protect public safety and reassurance that the situation is being addressed? Effective, timely communication — around the clock — is essential to keeping a city functioning.

- What do your mutual assistance agreements set in motion — between neighboring jurisdictions and different levels of government? They should be updated to be automatic at different levels of response, so that no time is wasted negotiating specific actions.

- The CSX fire had one more complicating factor, which is probably not atypical of a potential terrorist attack. We were forced to work in multiple locations: Camden Yards at the south end of the tunnel, Mt. Royal at the north end, and a manhole in the middle of downtown that was directly above the burning train.

- It is important to have an incident commander at each site — someone needs to have the authority to make an immediate decision, if needed, and take responsibility for safety. And there should be personnel from each participating agency to ensure effective coordination between each site and the central command center.

- After five days, the fire was extinguished, and the chemicals were removed without any serious injuries. Although our firefighters clearly demonstrated that they had been effectively trained, there were instances where we felt our response should have been more tightly scripted. Right away, we set about updating all of our emergency response plans.

- As a result of the train fire, we were better prepared on September 11th. We knew where to go. Our Police Department was the lead agency because the primary threat was to public safety. And we were better equipped to make decisions and disseminate information to the public.

- But at the same time, watching in horror as first New York and then Washington came under attack, we realized how much we had to do — and how much citizens depend on their local government to protect them if, God forbid, the worst should happen.

- The level of preparation and bravery demonstrated in New York was truly awesome. A complex where 50,000 people worked was essentially wiped from the face of the earth in a matter of minutes. About 6,000 people were killed, but 44,000 people were safely evacuated. Mayor Giuliani and his people deserve our awe and admiration.

- In the days following September 11th, Commissioner Norris and the Hart-Rudman Commission Report provided me with an understanding of what it would take to achieve the level of readiness present in New York on the day of the attack. But the worst-case scenario was suddenly much worse. And cities will need to achieve a previously unimagined level of preparedness.

- As all of you are demonstrating today by holding this hearing, we have a responsibility and duty to do everything we can to keep cities safe. Today, every big city mayor in America has a choice to make: Their city can be a hard target or a soft target, in the event of a terrorist attack. Baltimore is quickly becoming a hard target. And I'm glad to be here today, because I am trying to share what we are learning with anyone in a position to help. Obviously, Congress can make a major difference.

- One of the first things we realized — based on our experience in the CSX tunnel fire — was that rail yards and tracks, filled with chemical tankers and munitions cars, represent one of our most vulnerable targets.

- There is virtually nothing keeping either a terrorist or a lone kook from walking up to a toxic chemical tanker and blowing it up, releasing deadly gas into the air. There are no fences. There is no real security. There aren't cameras in tunnels.

- Baltimore is like every other city on the East Coast in this vulnerability. Hazardous materials are shipped by rail within yards of residential neighborhoods. There were large apartment buildings right next to the tunnel where our train fire burned. Commerce cannot stop, but this is one area where the federal government can make a significant impact.

- Local and State governments have limited ability to influence interstate carriers like CSX and Norfolk Southern to increase their security measures. Today, our railroads are an inviting target, and it doesn't have to be the case. Please push them to do more.

- We also have taken a series of steps to improve our preparedness within the past few weeks — with the goal of achieving a push-button level of readiness, where one event will trigger a chain of automatic responses from government agencies and private sector partners. I will leave most of the Police Department actions to Commissioner Norris, but since September 11th, we are focusing our efforts on three fronts:

1) Security:

- Holding daily security briefings with Police, Health, Fire, Public Works, Transportation and Information Technology Departments and State officials.

- Securing and protecting City's vulnerabilities, such as major buildings, water system, stadiums, major rail and interstate highway bridges and tunnels.

- Bolstering police and security presence at City buildings, including at water and wastewater facilities.

- Issuing an Executive Order requiring all City employees to display picture identification in all City buildings.

- Working with the chemical association of Baltimore to ensure that rail cars with dangerous hazardous materials are secured.

- Arresting and charging people who make bomb threats.

2) Emergency Preparedness:

- Consulting with a civil preparedness expert, former NYPD Chief Louis Anemone.

- Reviewing the findings of the Hart-Rudman Commission and its applicability to Baltimore — and consulting with Senator Hart.

- Coordinating closely with Center for Civilian Biodefense at Johns Hopkins University.

- Coordinating closely with U.S. Army Soldier and Biological Chemical Command ("SBCCOM") of the Department of Defense.

- Working closely with hospital CEO's on areas of preparedness and data collections.

- Developing transportation plans for road closures.

- Drafting standard operating procedures for each agency.

- Training personnel, including those who work in our water and wastewater facilities.

- Collecting and reviewing Baltimore City's existing mutual aid agreements with surrounding counties.

- Standardizing security and preparedness response with the State, as well as drafting upcoming legislation to create a statewide mutual aid agreement.

- Meeting with all public information officers, internally and from hospitals, to address communication issues and develop a communications plan.

- Meeting with press to discuss City's ongoing preparedness and dissemination of information in the event of an emergency.

3) Intelligence:

- Creating a web-based surveillance system to provide real time reporting from hospitals and ambulances, regarding infectious disease data and hospital bed availability.

- Testing reservoirs and the water system several times daily.

- Developing a statewide security intelligence network, working with other law enforcement agencies. As you may recall, one of the terrorist cells was based in Laurel, just a few miles south of our city.

- Meeting daily with Federal authorities to obtain intelligence.

This third area—intelligence—is where I have some concerns. The Federal Bureau of Investigation must recognize that effective police departments are a significant asset. The FBI has yet to ask our Police Department to follow-up on a single lead or tip. There are about 12,000 FBI agents, and they've received more than 100,000 tips. To me, this suggests that they are simply focusing on what they consider to be the "hottest" leads and taking their chances with the rest. Considering what we missed before September 11th, this is not a comforting thought.

Commissioner Norris has repeatedly offered our assistance — not just because of patriotism, but because we want to make sure our people are safe. But the FBI has yet to approach our Police Department to assist them with even the most basic of tasks, such as tracking down unreturned rental cars, or investigating people who recently have obtained pilots licenses, hazardous materials licenses or flight training. Law enforcement cooperation is not nearly what it should be, given what is at stake.

The steps I have outlined are just the beginning. Many of our preparedness efforts simply require making the time to complete tasks. But some of it is expensive. We do not have nearly enough equipment to protect our emergency response personnel. Communications equipment is necessary and very costly. And in the event of an emergency, our overtime costs will go through the roof, draining resources from other pressing needs.

While some functions are logically federal in nature (for example, maintaining a vaccine stockpile) other things can best be handled at the local level — if we are provided with adequate resources. The federal government must help put local governments in a position to succeed. Emergency preparedness cannot become another unfunded mandate, or eventually it will become unfunded, as priorities shift and dollars are spent elsewhere.

Moving forward, I believe the Chemical Warfare Improved Response Program operated under the most effective model. Given that so much emphasis must be placed on first response — which is almost always a local responsibility — funding for equipment and other priorities should go directly to local governments. This is what happened under the CWIRP, before funding was reduced. Passing these funds through State government would only result in implementation delays and, possibly, the diversion of funds for unnecessary administrative costs.

Life in the United States is different than it was just a few short weeks ago. You are right to focus on how we can best cooperate to rise to the challenge of our day. Although we all are certain our nation will prevail, these are uncertain times. And the most we can do... The best way we can protect the people we are privileged to serve is by eliminating as much of that uncertainty as possible through preparation.

In Baltimore, we are taking responsibility for doing as much as we can. We are not waiting for Annapolis. We are not waiting for Washington. That is the American way — neighbors take care of each other. If our city waited for advice on self-defense from Washington in the war of 1812, all of us would be singing "God Save the Queen."

However, that said, I am grateful that you are devoting your energies to this topic. Fighting terrorism and safeguarding our citizens is a National issue—it is a National challenge. And it will require our National resources to do all that we know can be done.

Thank you for the opportunity to testify here today. I am glad to answer any questions you might have.

End Testimony.

Appendix B

July 18, 2001

Thick, black smoke billows out of the railroad tunnel near Oriole Park at Camden Yards. Interstate 395 and the baseball park were closed, along with the Inner Harbor.

(Courtesy of the Baltimore City Fire Department)

July 21, 2001

A Baltimore fire commander emerges from a manhole near the intersection of Howard and Lombard streets after a stint in the Howard Street Tunnel which was filled with smoke from the CSX train fire.

(Courtesy of the Baltimore City Fire Department)

July 22, 2001

Baltimore City firefighters try to extinguish a burning rail car that was removed from the Howard Street Tunnel.

(Courtesy of the Baltimore City Fire Department)

July 18, 2001

CSX officials inspect two cars in the blackened Howard Street Tunnel.

(Courtesy of the Baltimore City Fire Department)

July 23, 2001

A firefighter hooks a cable to the door of the last railroad car pulled from the Howard Street Tunnel. The door was pried open so firefighters could extinguish the computer paper smoldering stubbornly inside.

(Courtesy of the Baltimore City Fire Department)